Julian Williams · Wolff-Michael Roth
David Swanson · Brian Doig
Susie Groves · Michael Omuvwie
Rita Borromeo Ferri · Nicholas Mousoulides

Interdisciplinary Mathematics Education

A State of the Art

Julian Williams
Manchester Institute of Education
The University of Manchester
Manchester
UK

Susie Groves
Faculty of Arts and Education
Deakin University
Victoria, VIC
Australia

Wolff-Michael Roth
Faculty of Education
University of Victoria
Victoria, BC
Canada

Michael Omuvwie
Manchester Institute of Education
The University of Manchester
Manchester
UK

David Swanson
Manchester Institute of Education
The University of Manchester
Manchester
UK

Rita Borromeo Ferri
Department of Mathematics and Natural Science
Universität Kassel
Kassel, Hessen
Germany

Brian Doig
Faculty of Arts and Education
Deakin University
Victoria, VIC
Australia

Nicholas Mousoulides
University of Nicosia
Nicosia
Cyprus

DOI 10.1007/978-3-319-42267-1

Library of Congress Control Number: 2016944914

Main Topics You Can Find in This "ICME-13 Topical Survey"

- Theorising and conceptualising discipline and Interdisciplinarity in Mathematics Education (IdME);
- Surveying and reviewing the empirical literature in the field of IdME; and
- Exemplifying Research and Development of IdME through case studies.

Contents

Interdisciplinary Mathematics Education: A State of the Art 1
1 Introduction 1
2 Survey on State of the Art in Interdisciplinarity and Interdisciplinary Mathematics Education. 2
2.1 Introduction 2
2.2 Interdisciplinarity: Historical and Theoretical Grounding. 3
2.3 A Survey of the Field of Empirical Research 13
2.4 Interdisciplinary Teaching and Learning in School—Case Studies. 21
3 Summary and Looking Ahead. 31
References. 32

Interdisciplinary Mathematics Education: A State of the Art

1 Introduction

This monograph has been written to provide a State of the Art in Interdisciplinary Mathematics Education (IdME), a relatively new field of research in mathematics education, but one that is becoming increasingly prominent internationally because of the political agenda around Science Technology, Engineering and Mathematics (STEM). In almost all countries now politicians see education in terms of preparation of a workforce for a competitive industrial sector, and STEM is seen as the route to more value-adding industries, especially in knowledge economies. Indeed in many 'advanced' countries there is almost panic at the prospect of declining numbers of qualified engineers and technologists leaving universities to these professions.

However, it is not only in the mainstream sciences that make up STEM that concerns are raised: in the social sciences too the professional and learned societies are expressing concerns at the lack of adequately numerate recruits. To illustrate, the subject of 'statistics anxiety' among non-STEM humanities and social science students has become prominent in the UK (see for example Onwuegbuzie and Wilson 2003). It seems as professional work becomes more mathematical that we will increasingly need to refer to 'mathematically-demanding' programmes and courses rather than just to those in STEM.

Consequently, the task of thinking about mathematics education in this context leads to an increasing concern for how mathematics inter-relates with the other disciplines and contexts involved: for most of the students of concern may only study mathematics for the sake of other 'leading' interests and activities, and they may even disidentify with mathematics. On the other hand if the interdisciplinary significance of mathematics can be understood, there is an opportunity in fact to encourage such students to reconsider and even revisit mathematics. Thus, 'interdisciplinarity' should be a major topic for mathematics education in particular, and

J. Williams et al., *Interdisciplinary Mathematics Education*,
ICME-13 Topical Surveys, DOI 10.1007/978-3-319-42267-1_1

we can expect it to become much more prominent in educational research and practice. (See e.g. Howes et al. 2013.)

In considering the main thrust of what needs to be gathered together in this State of the Art, we were struck by the fact that few in mathematics education have dealt with the theoretical and conceptual task of interdisciplinarity, and indeed few discuss theoretically the origins and social formation of the 'disciplines' that mathematics interacts with. This is a critical question: We need to know what we mean by the various forms of interdisciplinarity and curriculum integration, why it is being promoted and how it should be understood. Indeed, first we need to understand the concept of 'discipline', where the disciplines come from and how they inter relate. This is the first task of the survey.

Then we need an up to date, rigorous review of the empirical research literature on the topic: we need to know what has been done so our research can build on it. In the next section we report some progress towards this: we 'review the reviews', indicate the scope of the vast literature that searches throw up, and we illustrate the types of work in the literature. A full synthesis of the whole field awaits further work, however, and was beyond the scope of this book.

Finally we need to understand the kinds of interdisciplinary work being developed by researchers with practitioners in schools in their institutional and political contexts. The growing perception of a need for inter-disciplinary, practical, 'real world' problem solving has led to many often small scale initiatives that bring teachers together in an experimental project to develop the curriculum. Sometimes quite significant projects get funded on a large scale. We report two such cases that we have recently been engaging in. These are typical of many studies found in the literature: They deal with (one small scale, one large scale) opportunities to enrich learning experiences including mathematics (one STEM, one not), but also describe some of the demands of trying to develop an interdisciplinary profession (one in Primary, one in secondary schooling).

2 Survey on State of the Art in Interdisciplinarity and Interdisciplinary Mathematics Education

2.1 Introduction

As we have argued above, our reading of the literature in mathematics education suggests the need to clarify conceptually what is involved in the notion of a 'discipline', and so interdisciplinary work. Even consistency of terminology has not been established, but we mean more than this. How have 'disciplines' come about, what is at stake, why are the boundaries between disciplines notoriously difficult to cross, why interdisciplinarity is praised rhetorically but often so difficult to practice, and so on? We will argue that a social, historical account is necessary, one that explains how disciplines have become both socially functional and yet also dysfunctional.

Then there is a literature in educational research in general that has addressed notions of interdisciplinary work, especially 'thematic' work in Primary education, and integrated curricula in middle and secondary schools, and 'interdisciplinary' work in universities and beyond. We need to understand the state of the literature here, especially the different sorts of empirical studies that have been attempted, and how this can inform future research.

Finally there is 'practice': we need to understand what is happening in the field, how it is informed by research and how it can inform research. There is a vast gray literature and professional literature now essentially reporting curriculum development efforts. Some of these involve evaluations, often to satisfy funders who have made the interdisciplinary projects possible, but some of which can claim to involve evaluation research, as 'evaluation case study research'.

We therefore pose three research questions:

- What do we mean by, and how shall we theorise 'discipline' and 'interdisciplinarity' in mathematics education?
- What is known in the extant literature about interdisciplinary or integrated work across the disciplines in education?
- What is the state of the art in educational practice?

In addressing these questions, the following is divided into three sections, involving: the theory and conceptualization of interdisciplinarity; a survey of the empirical literature on interdisciplinary mathematics education (IdME); and case studies of interdisciplinary working in schools.

2.2 *Interdisciplinarity: Historical and Theoretical Grounding*

> All the human sciences interlock and can always be used to interpret one another: their frontiers become blurred, intermediary and composite disciplines multiply endlessly, and in the end their proper object may even disappear altogether (Foucault 1970, p. 357).

2.2.1 Introduction

The problem of interdisciplinarity requires an understanding of the concept of 'discipline' or 'disciplinarity'. Although there is an ongoing debate that holds classical disciplinarity to be extinct, the fact of continuing discussion of interdisciplinarity as a topic marks the problem as one that continues to be alive (Marcovich and Shinn 2011). In fact, disciplinarity may be understood as a multifaceted and nested system, where different forms of inquiry are situated at one or another level of complexity of the inquiry process: mono → multi → inter → trans → meta disciplinarity. Here it is suggested that 'inter' involves some sort of hybridising of the 'multi' disciplines (perhaps when chemistry and biology become biochemistry) while 'trans' implies transcendence due some sort of subsumption of the disciplines within a joint problem solving enterprise (perhaps when a new form of mathematics develops to deal with a problem such as calculating odds in gambling). Finally, in meta-disciplinarity, one becomes aware of the root disciplines in their relation and difference, e.g. when the nature of 'using evidence' in history and in science becomes contrasted but thereby clearer.

But then one comes to the notion of 'disciplinarity' in the professional world outside of 'science' proper: for instance one may speak of multi-disciplinary teams in the health service. Here the disciplines may appear simply in different job titles and remits, such as physiotherapist, nurse, teacher, general practitioner, and consultant. In this out-of-school context one sees many of the same issues arising in joint work as one does within academe and science: but now team work, professional or disciplinary 'identity' and division of labour are absolutely of the essence and must somehow be subsumed in the holistic interest of the 'health of the patient'. Each 'discipline' then has some sort of professional identity at stake, but must also prove itself as efficacious in the larger good, in the 'joint enterprise' or activity of health care.

Professional disciplines also often have their scholarly as well as practical 'knowledge bases' too, though their professionalism may be defined perhaps more often by practical competence than by their formal curriculum or scientific societies as such. Indeed many of these professional disciplines have spawned schools in academia, as they demand professional qualifications and accreditation: schools of engineering, nursing, social work, film, computer games etc. now being commonplace in universities. In this section then we illuminate both these sorts of disciplines in a general theory or conceptual framework of disciplinarity.

2.2.2 Disciplinarity

As argued above, we must begin the journey by seeking to understand how 'disciplines' arise and continue to flourish and even reproduce; and how they work separately and together to service social functions. Only then can we understand the difficulties and constraints—but also the opportunities—that interdisciplinary work poses. Disciplinarity is both (a) a phenomenon of the social world marked by increasing specialization and differentiation of (material and discursive) practices and (b) a form of discourse making the specialization thematic. Although the division of labour preceded the birth of the term discipline, the two aspects of disciplinarity have become intertwined.

Origins of the term. The English etymology of the term discipline points to the French language, where, in the 11th century, the term was used for punishment and pain. The term derives from the Latin *discipulus*, student, and *disciplina* (discipulina), meaning: teaching, instruction, training, branch of study, philosophical school, monastic rule, and chastisement (6th century). The Latin terms become the etymons for the French use of *discipline* during the Middle Ages, where it leads to the series of uses as 'massacre', 'carnage', 'teaching instruction' (12th century), '(body of) rules of conduct' (12th), 'punishment', 'self-control' (12th), 'branch of learning' (14th), and 'knowledge of military matters' (beginning 15th). In English, "discipline' was used in Chaucer's time to refer to branches of knowledge, especially to medicine, law, and theology, the 'higher faculties' of the new university' (Shumway and Messer-Davidow 1991, p. 202). In the sociology of knowledge, the origin of culture and social representations has been situated in the religious forms of life, with its own rites, discipline of the body, and asceticism; science, which took the place of religion, nevertheless is characterized by these forms characteristic of the religion it replaced (Durkheim 1915).

Definition. A discipline may be defined as 'a specialized pursuit of circumscribed scope' (Mannheim 1956, p. 18). Although a discipline previously had been characterized in terms of association and differentiation (Simmel 1890), the concept in itself does not capture the phenomenon as a whole. There is also the object of inquiry and the system of shared significations. Acceptability of a new discipline was brought about and thought in terms of a 'hallowed principle of specialization at any price' (Mannheim 1956, p. 19). Disciplines are generated through the focus of inquiry or work, which, as its reverse, may lead 'to voluntary blindness to problems which straddle the agreed borders of two or more disciplines' (p. 20). The foci or objects of inquiry and work are associated with discursive practices, 'groups of statements… that tend to coherence and demonstrativity, which are accepted, institutionalized, transmitted, and *sometimes* taught as science' (Foucault 1972, p. 178). The rules and procedures operative in scientific investigations—material and associated discursive practices—are specific to the discipline, in particular in those situations where there is a recognition that they have to be appropriate to the object (Bourdieu 1992).

Unit of analysis. In a (Marxist) sociological account, disciplinarity is treated as a social phenomenon. In sociology, society not only is taken to be a phenomenon sui generis but also the phenomenon that distinguishes humans from other species (Durkheim 1915; Marx and Engels 1978). The smallest unit of analysis for any *specifically human* phenomenon, therefore, has to be one that has all the characteristics of society as a whole. One such unit is 'productive activity', involving the production of things for consumption, i.e. meeting human needs. Productive activity, including its particular distinct material and discursive practices, together with needed consumable products, can be seen as defining a discipline. Arguably, then, disciplinarity did not exist from the beginning of humankind, but came into being as 'disciplined activity', to meet some need. The many different forms of production that exist today historically have emerged as a result of increasing division of labour, specialization, and 'out-sourcing', and in some specialities an infrastructure of 'teaching' characterizes it as an emerging discipline. That is, in the course of history, the nature of a discipline changes—what used to be philosophy (the production of knowledge of truth, e.g. during ancient Greece) later bifurcated into philosophy and physics. Society as a whole can be thought in terms of the ensemble of interconnected societal activities; and the division of labour is a principal force for the cohesion of society (Durkheim 1893) and a source of its inner contradictions (Marx and Engels 1962). By participating in an activity, that is, by contributing to the generalized production of goods and control over human conditions to meet needs, individuals increase control over their individual conditions and needs satisfaction. The interconnection occurs by means of exchanges, whereby the products of one activity become an integral part of another activity, as when the production of new knowledge within a discipline like mathematics becomes part of the tool kit for associated productive activities in engineering or manufacture. The category of activity, thereby, includes all the aspects that traditionally are attributed to a discipline with its characteristic community of practitioners.

Each productive activity involves a community in collective, joint labour: it may be characterized by a dialectical unity of a number of moments, including the *subjects* and *objects* of activity, but significantly 'mediated' by the whole historically-evolved system of production (involving means such as tools and signs, conventions and rules, and the reigning division of labour, see Engeström's schema in Fig. 1). Most importantly, subjects' activity also is dialectically both motivated by and causative of the subjects' *consciousness* and *personality* (Leont'ev 1978). It is impossible to understand the relationship between discipline and institutions if we fail to acknowledge their basis in productive activity and its historically-produced mediating conditions, which explain power relations and oppression for instance (Bourdieu 2000).

2.2.3 History of Disciplinary Nature of Human Praxis

Understanding the cultural historical legacy that is entailed in our 'disciplines' may help us to understand the nature of the disciplines themselves. But it also may help us understand why inter-disciplinary work can be difficult, confronting certain sorts of obstacles, power structures, and questions of identity, differences in understandings of knowledge, discourse and practice.

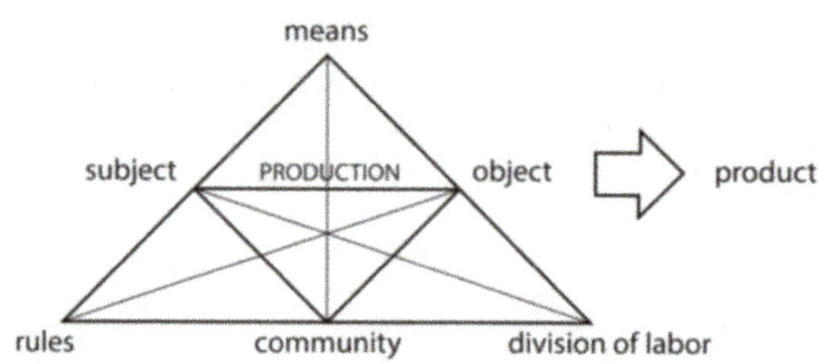

Fig. 1 Cultural-historical activity theoretic formulation of societal activity, the smallest unit that has all the characteristics of society (after Engestrom)

Dawn of the disciplines. In classical sociological approaches in the Eurocentric tradition, formal notions of discipline and formal aggregations around particular practices are said to have emerged at the beginning of the Middle Ages, their origin dates back to ancient Greece with the emergence of industries besides agriculture (Durkheim 1893), involving inter-city, inter-state and even international divisions of labour and trade. Discipline as such requires a form of corporation in an institutional form ('community' in Fig. 1), for an aggregate of people does not in itself constitute a discipline. During the Roman Empire, the different trades came to be treated as entities with particular functions in the public service, the charge and responsibility for which lay with the corporation. Because the service was imposed, requiring state sanctions to maintain it, the corporations ceased to exist with the end of the empire. In the European context, they were reborn in virtually all societies during the 11th and 12th centuries, when tradespeople felt the need to unite, forming the first confraternities.

Confraternities as disciplinary organizations. The confraternities and the guilds they gave rise to, as authorities regulating the practices of their members ('rules' in Fig. 1), can be seen as the first organizational structures that exert themselves as forces on the formation of the durable dispositions of its members. Such regulation occurs 'through all the constraints and disciplines that [the organizational structure] imposes uniformly on all agents' (Bourdieu 2000, p. 175). In the European context, the training of traditional artisans began with apprenticeship, which ended when aspiring individuals became journeymen upon successful completion of a specific piece of work in and with which they exhibited specific skills. As journeymen, they literally traveled and worked in different locales until ready to complete a 'master piece' to be judged by members of the guild. Through the masterpiece, journeymen exhibited mastery of the means of production (Fig. 1) and the form of consciousness required for the transformation of objects into a craft-specific product. If successful, they became masters and obtained the right to have their own shop, train apprentices, and employ journeymen. The old forms of reproduction were reborn in the division of training and work, cross cut by another division of theory and praxis, the former occurring in (vocational) school and college, the latter as practical apprenticeship or 'experiential learning'. Even the designation of 'masters' found a new life in the 'Masters degree', and the trade certificates mutated into high school and college/university diploma.

Separation of theory from practice. The increasing division of labour partially is the result of the increasingly specialized knowledge required to do a particular job. 'The production of ideas, of conceptions, of consciousness, is at first directly interwoven with the material activity and the material intercourse of men—the language of real life' (Marx and Engels 1978, p. 26). This same progressive division of labour also split theory and practice, the former often being taught in schools, the latter on the job. Indeed, 'division of labour only becomes truly such from the moment when a division of material and mental labour appears' (ibid, p. 31). For example, large constructions prior to the Gothic era were organized by master masons, who directed the work routine. It is out of this occupation that the division into architects and labourers emerges (Turnbull 1993). Architecture itself subsequently differentiated into disciplinary forms emphasizing design, on the one hand, and engineering, on the other hand. In the history of intellectual (theoretical) disciplines, 'the specificity of the scientific field stems from the fact that the competitors agree on the principles of verification of conformity to the "real", common methods for validating theses and hypotheses' (Bourdieu 2000, p. 113). Numerous case studies show how new disciplines or non-disciplinary fields—penology, education, nursing, midwifery, biology, or psychiatry—are tied to specific, shared discourses and practices; economies of concepts; supporting institutions; conditions and procedures of (social) inclusion and exclusion; transmission and training; relations to law, labour, and morality; and (disciplinary) practices or technologies of surveillance, government, and control (Foucault 1970, 1978, 1988). Archaeological, genealogical and critical studies together also exhibit who controls existing discourses and how these constitute the very boundaries of a new discipline. As a result, a focus on 'disciplinary boundaries' rather than 'discipline' can help reveal an understanding of the phenomenon as a combination of internal and external social processes (Fuller 1991).

Origin of academic disciplines. The academic, scholastic disciplines have their Western origin in the medieval divisions of the trivium (grammar, rhetoric, logic) and quadrivium (arithmetic, geometry, astronomy, music) that lasted to early modernity (d'Ambrosio 1990). The sciences originated in philosophy, 'which fragmented itself into a multitude of special disciplines of which each has its object, its method, its mind' (Durkheim 1893, p. 2). The objects of inquiry and the principles on which they are based historically were re-ordered towards the end of the 18th century and the arrival of mathematization. Before Kant's critique of reason, representations were inherently linked. With *mathesis*—i.e., the systemizing practices establishing the order of things—an epistemological differentiation occurred, according to archeological and genealogical analyses, into a field of 'a priori sciences, pure formal sciences, deductive sciences based on logic and mathematics' and a field of 'a posteriori sciences, empirical sciences, which employ the deductive forms only in fragments and in strictly localized regions' (Foucault 1970, p. 245).

Societal function of discipline. In sum, a discipline functions as 'a system of control in the production of *discourse*, fixing its limits through the action of an identity taking the form of a permanent reactivation of the rules' (Foucault 1972, p. 224). One cannot speak 'the truth' outside of such a system, as can be seen in the case of 19th century biology, where the statements of Gregor Mendel about heredity made no sense to contemporaries. It was only after a complete shift in the disciplinary discourse of biology itself that Mendel's statements, its objects and discourse, were recognized as true. That is, one can 'only be in the true… if one obeyed the rules of some discursive "policy" which would have to be reactivated every time one spoke' (p. 224). In this analysis, (disciplinary) forms of discourse, though also an opportunity, first of all need to be thought of as a constraint. This constraint arises in part from the acceptable forms of representations and the associated practices that both constitute and distinguish the discipline and its boundaries (e.g. Hine 1995; Lynch 1985).

2.2.4 Physical Discipline and Forms of Thought and Practice

From the definition of *discipline*, it is apparent that the term constitutes a double-edged sword: (a) it specifies the organized ways in which scientists and practitioners go about their work such that they can indeed be identified in terms of specific practices; and (b) getting to the point of exhibiting these practices requires physical and mental discipline, generally instilled by imposing (more or less severe) constraints in the way persons work.

Historical associations. Discipline is a historical product in all its senses and connotations. Even military discipline, today the epitome of discipline, was the result of a historical development towards physical, material, and behavioral standardization. These disciplinary forms, to achieve cohesion and esprit de corps, are but the limiting case of disciplinary training (Bourdieu 2000). The emergence of discipline in the military thus falls together with the emergence of formal (mass) schooling, which, through its practices, constituted not only physical and mental discipline but also a system of social ordering (Foucault 1978). The combination of both physical and intellectual dimension emerged in the Church, where the first 'well-amalgamated and disciplined intelligentsia' (Mannheim 1956, p. 130) was born. Strict adherence to specified higher linguistic forms constituted a self-imposed discipline that required exercise and physical discipline for its achievement.

From physical to mental discipline. The role of physical hardships in the emergence of (a) discipline already had emerged with the recognition that the discipline required for working in factories was instituted by means of bloody legislation—beating vagabonds bloody, cutting of part of the ears, and the death penalty constituted a discipline subjecting 'free workers' to slave labour (Marx and Engels 1962). The coercive aspects of disciplines were exhibited in historical analyses of pedagogy, penology, psychiatry, and clinical medicine, which emerged as 'science discourses', according to Foucault (1978), by means of an institution of networks of files, accounting books, timetables, drill exercises, and their associated practices. But the emergence of discipline occurred not in the abstract. Instead, 'the success of disciplinary power derives no doubt from the use of simple instruments: hierarchical observation, normalizing judgment and their combination in a procedure that is specific to it, the examination' (p. 170). Recent sociological and anthropological studies show that in many disciplines, a period of physical and emotional hardship is an integral part of the intellectual trajectory of a person in the process of becoming recognized as a member of an elite field of inquiry (e.g. Delamont et al. 2000). Thus, for example, the graduate experience in ecology may involve repeated, lengthy stays in isolated areas where candidates do fieldwork, which may involve exposure to inclement (extreme) weather, physically exhausting data collection, and isolation (Roth and Bowen 2001). This association of discipline and the particular cognitive regimes is not new, but has its origin in religious practices of asceticism that developed in the monastic orders and elsewhere.

From exacted discipline to self-discipline. The self-discipline that characterizes the member of a discipline might include: spatial organization of desks and lectern, temporal organization of the school day, section of the curriculum, systems of punishments and rewards, specialism of teachers, and geography of the school buildings, all contributing to disciplining the mind through the disciplining of the body and of associated discourses. Differential success in the system, measured in terms of points, led to differential access to different ranks in the military (Foucault 1978). Others, too, refer to the 'mechanisms of election', which 'leads the elect to elect the School which has elected them, to recognize the criteria of election which have constituted them as elite' (Bourdieu 2000, p. 35). Ultimately, we find self-discipline again as an organizational force at the institutional level, for example, in peer review and the self-managed academic faculties (dean, doyen, primus inter pares). Self-discipline therefore is an internalized form of the external discipline demanded through controlled, controlling systems that promise order at the societal level. According to Foucault, the various social sciences disciplines not only reproduce themselves through the microphysics of power–knowledge in formal institutions but also, in their field of inquiry, they constitute forms of social control.

2.2.5 Working Across Disciplines

All spiritual practices, even before the arrival of epistemology, have recognized the relationship between the quality of practice and thought and the extent of training and practical experience. However, the affordances associated with increasing disciplinarity are accompanied by the above-noted blindness. From this blindness are 'distortions in relations with the representatives of other disciplines' (Bourdieu 2000, p. 176)—i.e. the very phenomena that interdisciplinarity has to address. This creates obstacles to working in an interdisciplinary manner.

Common objects as conditions. The problem of interdisciplinarity may be framed in terms of the activity theoretic approach outlined above. Here, paradigmatically, in contrast to the normal organization in society, whereby products are exchanged by means of a generalized exchange form, i.e. money, two or more groups (organizations) representing different disciplines, may come together to work on a common object but with no medium of exchange. Thus, for example, one study reported how an interdisciplinary project emerged when three 'relatively autonomous project groups, composed of researchers with different disciplinary backgrounds' came together for the purposes of constructing 'the key parts of this projected production system: the development of microbial strains' (Miettinen 1998, p. 430).

A note here is needed to understand how Activity Theory conceptualizes the 'object' or better 'object/motive' of activity. An activity is defined by the 'object' on which the collective is working, but the object (like all the 'moments' represented as nodes in Figs. 1 and 2) is in the process of transformation: the work of the collective involves transforming the 'raw' objects into 'outcome' objects that meet social needs. In the classical case of labour activity, the actions of the various workers lead to the manufacture of finished consumables. Thus the 'motive' involved is the envisaged transformation of the 'raw' object into 'outcome object',

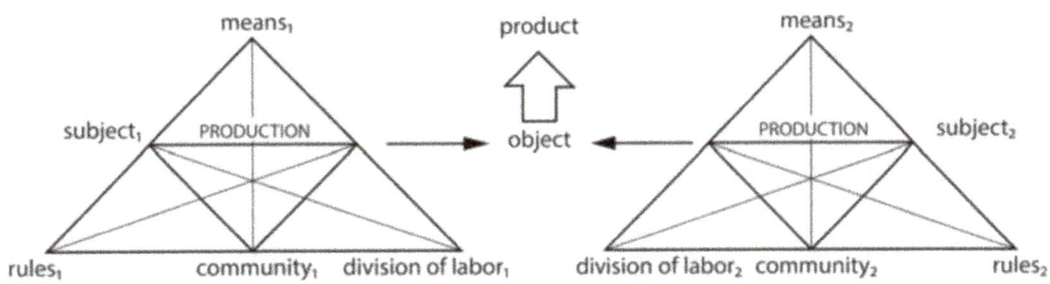

Fig. 2 In an interdisciplinary project, two different activity systems collabourate with a common object/motive

sometimes this motive is imagined, sometimes only emergent through the collective actions of the individual subjects involved. When we speak of 'object/motive', then, we have all this in mind: an activity is defined by the 'motive' of transforming an 'object' into a new form that meets a social need.

Thus, common objects (object/motives) often characterize interdisciplinary projects, even though the contributing disciplinary activity systems differ, each with its own distinctive characteristics, tools, and perspectives (Fig. 2). The possible contradiction is immediately apparent. Because each part of an activity system is a function of the whole and is perfused by the characteristics of all other parts, the motives characterizing any two activity systems may differ. That is, any interdisciplinary endeavor involves the work of specifying a *common* object-motive (product), which likely differs from object-motive$_1$ and object-motive$_2$ that characterize the respective mono-disciplinary efforts.

The difficulty in defining a common object can often explain the failure of projects designed to be interdisciplinary. On the other hand, in successful projects, new objects/motives are created in such a way that they make sense within each of the disciplines (e.g. Miettinen 1998). A good example of such an endeavor was observed in the collabouration of printers and designers to redesign the printers' workplace (Ehn and Kyng 1991). Together, representatives from the two disciplines built mockups to model what happens in the workplace, and, in so doing, developed a new form of discourse that made sense within each discipline and constituted a sense-giving field that made sense across the fields.

Boundary objects and boundary crossers. One function of common objects (e.g. representational tools) is that they coordinate the activities involved even though the practices surrounding these objects differ. These objects are known as *boundary objects*: such objects define boundaries between practices (forms of activities). Thus, for example, in the manufacture of an aircraft, many different disciplines are involved; the coordination between these very different disciplinary fields is achieved by means of drawings (Henderson 1991). These drawings have different functions and are understood differently on the shop floor, in the accounting department, for the electrical engineers, or the inventory control department. Because of this, the object also may be thought of as a *conscription device*, that is, an entity that brings together (enrolls) members of different disciplines (communities of practice) for the purpose of realizing a common object/motive, which also is itself defined by that same device. There may also be small numbers of individuals who are familiar with and exhibit expertise in two disciplinary fields (Star 1995). They cross and transcend boundaries, sometimes being called 'brokers', 'wizards' or 'gurus' that are highly competent in multiple domains and across multiple systems of formal representations.

Stratification, hierarchies and dissent. Anyone working in academia knows about the institutional hierarchies between and within faculties and forms of knowledge. Thus, the natural ('hard') sciences tend to be regarded as higher in esteem and more powerful than the social ('soft') sciences (Bourdieu 2000); within a particular field, the same gradations are reproduced—e.g., in psychology, there are gradations from the 'hard' (e.g. experimental and physiological psychology) to 'soft' (counseling psychology); and within each field there are gradations, where some scholars are on top of the heap and others are mere 'foot soldiers'. The disciplinary divisions between hard and soft sciences found a parallel in gender divisions, which was the result of systemic institutional practices that systematically excluded women from the natural sciences (Shumway and Messer-Davidow 1991). One of the conditions for interdisciplinarity to emerge is the active engagement with the historically developed attitudes between disciplines and forms of inquiry for the purpose of overcoming divides set up by condescending and colonizing attitudes. In the project of interdisciplinarity, one may as well heed the advice of someone who has studied the history of discipline: 'We must henceforth ask ourselves what language must be in order to structure in this way what is nevertheless not in itself either word or discourse, and in order to articulate itself on the pure forms of knowledge' (Foucault 1970, p. 381).

The structures of power gradations that separate the faculties and disciplines—while they have a degree of autonomy—are homologous with the entire field of power in society at large: natural sciences being opposed to faculties of social sciences (Bourdieu 1984). There is thus a totality of economic, cultural, and social differences. The ruling relations within disciplines reproduce those between faculties. Knowledge is a form of 'symbolic capital', a commodity that may be accumulated as any other form of capital. In the sociology of symbolic capital, the university faculties are characterized by their position within the academic field of power, each with its own internal field of power and cultural capital. Within disciplines, certain schools, sometimes associated with specific universities (e.g. Ivy League) reproduce these structures through filiation and graduate student exchange (e.g. Traweek 1988). Hierarchical relations result because some disciplines have more fundamental or universal topics and applications, such as the sociology of mind, 'in as much as social situations are tacit components of all mental acts, no matter what academic disciplines or socially established divisions have custodial care of them' (Mannheim 1956, p. 54). Inter-faculty differences play out in a hierarchical system of power that gives differential access to resources. The practices of selection and indoctrination within each discipline contribute to the reproduction of differentiation between the disciplines. Cultural capital contributes to the constitution of a discipline within society as a whole and to the relative status of the individual within the discipline. This entire disciplinary formation therefore acts as a great weight on the world, making for difficulty in expecting those 'schooled' and 'disciplined' in one field to relate in effective ways with others whose habits have been formed in relatively independent, and contradictory fields.

A way forward—situating situated inquiries. Rather than making interdisciplinarity the new scientific dogma or ideal practice to be achieved, a more productive approach may consist in situating inquiries and endeavors according to the complexity of their questions, tools, objects, and outcomes. Thus, on the scale of complexity, interdisciplinarity may actually be thought of as a continuum of relations between disciplines, between mono-disciplinarity, on the one end, and meta-disciplinarity, on the other end, with multi-disciplinarity and trans-disciplinarity offering more or less hybridity of the disciplines involved between these extremes (Collen 2002). As a result, neither mono-disciplinarity nor any other 'level' is displaced; indeed the core value of the discipline may provide precisely the value to other disciplines that interdisciplinarity requires. But the continuum allows inquirers to advance by moving towards more complex inquiries involving more than one discipline in ways that lead to advances and novel forms of insights (e.g. Hicks 1992) or to return to less complex inquiries to draw on the advantages that arise from lower levels of complexity (in objects, organizational forms, efforts).

In this view, we conclude, interdisciplinary mathematics education offers mathematics to the wider world in the form of added value (e.g. in problem solving), but on the other hand also offers to mathematics the added value of the wider world.

2.3 A Survey of the Field of Empirical Research

2.3.1 Introduction and Caveats

Here we provide an outline of a systematic survey of literature related to interdisciplinarity and mathematics education, and a preliminary review of this literature. There are several limitations of and obstacles to our search and review: scope, scope and terminology. The first is scope in a general sense. All literature related to interdisciplinarity as such, and all literature related to the history of disciplines, school subjects and curriculum structure, even if not directly related to mathematics education, could have some relevance to our topic. In general, it can often be helpful to investigate the wider systems of practice, or of understanding, which an object of study is part of, because the object's relationships within the wider system mediate and structure it. Looking at these wider connected issues (and beyond those too) could therefore help us understand our particular object 'interdisciplinarity and mathematics education'. Although this is a general point, it seems particularly apt when discussing interdisciplinarity, a topic that in itself opens up challenges or questioning of narrow categorisations. For example, it has been suggested that 'specialization and expertise remain the coin of the academic realm, for reasons of ease of measurement rather than any inherent virtue to the approach'

(Frodeman 2010, p. xxxiv). Despite sympathy for this perspective, we here have restricted such wider perspectives to the overview given in Sect. 2.2, rather than expand our literature search to the almost infinite.

Then there is a second aspect of scope that also acts to confound our review. If the previous obstacle could be likened to infinity, the infinite which lies seemingly beyond our object of study but which actually influences or is part of it, then this obstacle is more analogous to the infinite density that lies between 0 and 1. The field of interdisciplinarity can loosely be seen as ranging from anything beyond a pure disciplinarity at one end of the spectrum, through to the complete dissolution of disciplines at the other end. Despite the preceding categorization of interdisciplinarity by level of complexity—monodisciplinary = 0, multidisciplinary = 1, interdisciplinary (now as a particular of the general term 'interdisciplinarity') = 2, transdisciplinary = 3, and metadisciplinary = 4—these categorisations themselves may be seen to be as open to questioning and doubt as the disciplines are themselves. For example, does mono-disciplinarity exist in its purest form? Even the most isolated academic mathematicians will do *something* outside of mathematical activity, which may influence their thinking within the mathematical world. Indeed, which mathematics counts as the 'discipline'? Is it the formal shell that remains when all connections to 'real' human activity have been removed, or is the mono-discipline the mathematics that retains the concrete within its abstractions? How do we differentiate an interdisciplinary approach to science that brings mathematics in as a tool, from that which brings in mathematics as a generalisation of scientific concepts? Where on our scale would we place a mathematics teacher conducting a peer observation of a drama class to develop their pedagogy? Would it require a fractal dimension? Such questions can be asked along all points of our continuum/scale, and hint also that there may be all sorts of relevant literature which may not self-define as being interdisciplinary.

The final obstacle to completeness is terminology. Although some have worked hard at clarifying the vast array of terms which interdisciplinarity may be known as and developing a shared language (e.g. Klein 2010), the literature itself does not necessarily conform to this logic and guidance. Also, given the preceding argument, even all the existing terminology taken collectively may not capture everything that we would wish it to.

Formal literature review. To review the literature, we followed systematic review guidelines that prioritise transparency of methodology (and so replicability), and quality. We were aware from the start that we would not have time to complete a comprehensive synthesis in the scope of this survey, but the aim was to be illustrative. In this case a ProQuest search was conducted within English language peer-reviewed journals classified as 'educational research' journals for (a) Interdisciplinary AND mathematics AND classroom; (b) Interdisciplinary AND 'mathematics education'; (c) Multidisciplin* OR Transdisciplin* AND 'mathematics education'; (d) 'integrated stem'; and (e) 'Integrated curriculum' AND mathematics.

This search led to 612 items (too long a list to include in our bibliography here: a spreadsheet is available on request of the authors). These were then supplemented through an ad-hoc process via their references and citations to give 754 items. A random sample of 250 of these texts was categorised in various ways to give an illustrative guide to the content and emphases found within the literature. This categorisation found that 80 % of texts were relevant to the review. Of these, 30 % were professional in nature, primarily outlining suggested integrated lesson materials. After removing these from the survey, it was found that 82 % included empirical data, the remainder dealing only with the idea, theory or concepts of interdisciplinarity.

In the following subsections, (a) we describe the results of our review of review studies, (b) we overview some of the large-scale studies, and (c) we exemplify the small-scale studies through some illustrative and typical examples.

2.3.2 Review of Reviews

We begin with the items that themselves claimed to be 'reviews of literature': these were not necessarily attempts to review the whole field, nor were they up to date. For instance one of these from the Review of Educational Research focused on just the meta-disciplinary aspects involving understanding the nature of different curriculum subjects (Stevens et al. 2005). We also drew on the Royal Society report that several of us had been involved in writing, and that included a literature search and review of curriculum integration and interdisciplinary work (i.e. Howes et al. 2013).

What emerges from an overview of already existing literature reviews is a range of common themes. Berlin and Lee (2005) offer a historical analysis of English language literature on the 'integration of maths and science education' through the 20th century, noting the widespread policy enthusiasm for curriculum integration in its last decade. They note two important trends (comparing the first nine decades with the last decade): (a) the growth in literature throughout main school education, and (b) the particular growth in work on secondary education, where integration has been seen as more demanding. Additionally they note the recent growth in literature articulating 'models' of integration, but that the empirical studies of these were 'weak'. A review of the U.S. middle school experience found little direct evidence of children's learning outcomes, as manifested in measures of student achievement or cognitive processes (St. Clair and Hough 1992). The majority of the studies reviewed focused on affect (student, teacher) and learning environment. Yet the review concluded 'that interdisciplinary curricula and instruction holds promise as a way of meeting middle grades students' developmental needs by making the subject matter relevant to real life and thus engaging them in the learning process' (p. 25). Czerniak et al. (1999) similarly bemoaned the dearth of empirical studies, pointing to a lack of consensus and clarity in what 'integration' means, and identified arguments for and against 'integrating' curricula in various ways.

Using mixed methodology, Hurley's (2001) review addressed some of these issues and 'found quantitative evidence favoring integration from a meta-analysis of 31 studies of student achievement, qualitative evidence revealing the existence of multiple forms of integration, and historical evidence of publishing patterns from across the 20th century' (p. 259). Their analysis of 31 studies reported positive effect sizes of five distinct 'forms' of integration on learning outcomes in mathematics and science, with highest effects being 'sequenced (on maths)' and 'enhanced (on science)'. Generally the outcomes for mathematics measures were less good than for science. This may demand a close analysis of these cases and what the 'forms of integration' involve, while acknowledging that almost any integration of maths with science would seem likely to benefit mathematical competences in science tasks while the reverse might not be so obviously the case. In fact, 'structure mapping theory' provides a cognitive explanation for this, as mathematics is seen as the abstract 'general' while science is seen as the 'rich context' in which the mathematics can provide structure, but on the other hand a context from which it may be cognitively demanding to abstract (Silk and Schunn 2011).

What has happened since 2001? In general these emergent themes remain the same. Becker and Park (2011) conducted a meta-analysis of twenty-eight studies, calculating thirty-three effect sizes, to address some key questions in relation to the integration of STEM subjects. The questions included those on the effect of integrative approaches- the variation of this effect in different year groups, the most effective approaches in terms of achievement, and the variation of achievement effects across different subjects. They found that integration at elementary level has the largest effect, as does integrating all four of S, T, E and M. They also found that the positive effects of integration were the smallest in relation to mathematics achievement, but argue that the increased student interest in the subject due to seeing its real-world connections, may lay the basis for improved achievement in the longer term. They also strike a note of caution for their meta-analysis given the 'very few empirical studies on the effects of integrative approaches' and stress the need for further research.

Honey et al.'s (2014) survey of integrated STEM decries the poorly described and poorly designed research which has occurred to this point, but nevertheless argues the potential in integrated approaches, alongside maintaining focus on individual subjects.

A review of curriculum architecture as a contribution to Scotland's 'Curriculum for Excellence' also suggests that the research on interdisciplinarity is inconclusive, though student engagement may be increased by the kinds of task that tend to be used in integrated curricula (Boyd et al. 2014). Whereas barriers are recognized, interdisciplinary approaches 'can offer opportunities for "joined-up" learning which subjects cannot always offer. They may also, paradoxically, help learners towards a clearer understanding of the contribution of individual disciplines' (p. 10). We refer to this as meta-disciplinarity, which suggests this 'paradox' of interdisciplinarity requires that learners acquire 'comparative understanding of school subjects', for example, that the logic of argumentation and use of evidence in History has some things in common, and some differences with its logic in Science, or in mathematics (Stevens et al. 2005). This then reveals a new 'learning outcome' that empirical studies have not actually measured, and one that might be thought educationally important.

There are various factors that can help enable, or hinder, attempts at implementation of interdisciplinarity or integrated curricula (Venville et al. 2012). These factors include 'subject matter knowledge, pedagogical content knowledge and beliefs… instructional practices… administrative policies, curriculum and testing constraints… school traditions… school organization, classroom structure, timetable, teacher qualifications, collaborative planning time and approach to assessment' (p. 737). Community expectations of more traditional forms of teaching may also be a factor. Venville et al. are explicit about the radical nature of curriculum integration that occurs with interdisciplinarity, and they argue that powerful forms of knowledge can arise for students through it. Their high level of emphasis on the obstacles to integrated work in practice flows from an active desire for change.

Obstacles to radical change also have been reported in the case of the failure of the expansion of integrated studies in Japan, which arose from the 'lack of proper investment in development, support, and infra-structure that might have facilitated genuine enactment of curriculum innovation' (Howes et al. 2013, p. 9). There is nevertheless a space in some curricula for problem-centred project work involving multi-disciplines, but this requires much support and space for teachers, and developments in assessment practice, for example. Again, a radical curriculum is seen to require radical change and support in wider domains (as is revealed in Sect. 2.4.2).

In sum, these existing literature reviews suggest collectively (a) a need for another up to date and more comprehensive review of literature; and (b) a relative dearth of systematic empirical work that builds cumulatively. Tentatively, however, they suggest that there is evidence of learning gains from integrated curricular and interdisciplinary working, mainly for learning outcomes of affect, of problem solving processes, and of metadisciplinarity. This might be an important qualification, as it suggests that the outcomes that will likely be affected by inter-disciplinary working will be non-traditional, and non-standard. Clearly if this is the case then studies that measure only traditional outcomes may find little 'positive' effect, and practices that are dominated by systems that value only traditional measures will likely swiftly reject IdME and integrated curricular approaches.

2.3.3 Large-Scale Empirical Studies

A number of large-scale studies have been conducted focusing on interdisciplinarity in mathematics education and we review these here. This is an eclectic mix of studies: we describe them first and then proceed to discuss their collective significance.

A project involving the enhancement of career and technical education with higher levels of embedded mathematics, aiming to develop mathematical understanding through teaching it in its 'natural context' was seen to have positive outcomes (Stone 2007). A total of 131 teachers in five curricular areas partnered with mathematics teachers to produce lessons for almost 3000 students. Findings showed statistically significant gains in traditional measures of mathematics skills.

In another project, middle school teachers of science and mathematics understandings of integration were surveyed via reflection on various presented scenarios, and description of their own attempts at integration (Stinson et al. 2009). Differences were identified in their characterisations of integration, and content knowledge was perceived as a barrier to integration.

Shulman and Armitage (2005) report on a five-year project where middle school teachers developed interdisciplinary discovery-orientated activities in workshops involving undergraduate students from a variety of subjects as teaching assistants. This led to a significant increase in students meeting required standards on standardized mathematics tests, and encouraged a number of the undergraduate students to pursue teaching careers.

Dorn et al. (2005) assessed GeoMath, an interdisciplinary unit of Geography and Mathematics introduced in order to combat the declining classroom time dedicated to geography teaching, especially in grades K–8 in the US. The GeoMath themed interdisciplinary unit consists of 80 lessons that were taught by 28 teachers in 113 pilot classrooms that mirror Arizona's diverse demographics. The individual activities include, amongst others 'Shape of My World: Mapping a Classroom', where 'students identify basic shapes in the classroom and make a map showing major furniture location and classroom features', and 'Counting islands: What is an island and how many do you see?', where 'students learn that the world is made up of many landforms, while practicing counting skills' (p. 154). The outcome of this study was that there were statistically significant increases in performance in students mathematics skills coupled with an improved understanding of geography standards. Also, the results show that 25% of the teachers involved in the teaching of mathematics reported increases in their confidence levels. In light of their findings, the authors suggested that there should be a national agenda of articulating the geography curriculum to high-stakes tested subjects of reading and mathematics. On the other hand, a study on interdisciplinary team teaching found no significant differences for reading, mathematics, science and social studies achievement (Alspaugh and Harting 1998). This was based on a study of the effects of themed interdisciplinary teaching as against single discipline teaching in middle schools. The scope of this study was limited.

Parr et al. (2009) conducted an experimental study involving teachers and students from 38 high schools in Oklahoma including 447 'Agriculture, power and technology' (APT) students (experimental n = 206; control n = 241). They posited that those students 'who participate in a contextualized, mathematics-enhanced high school APT curriculum and aligned instructional approach would develop a deeper and more sustained understanding of selected mathematics concepts than those students who participated in the traditional curriculum' (p. 59). The authors found that a mathematics-enhanced APT curriculum and aligned instructional approach did not result in a significant increase in student mathematics performance as measured by either conventional standardised mathematics tests or 'real-world' problem-based tests. However, implementation of the program was reported to be incomplete by some teachers, and the program did significantly affect a students' perceptions of the need for postsecondary mathematics education.

These studies are illustrative of the main large scale empirical studies in our literature sample, and exemplify the wide variation that appears in such studies in the literature. We see variation, for example, in who is measured (teachers or students; students in primary through to university); what is being integrated (e.g. which subjects); the nature of the interdisciplinarity involved (e.g. multi- or trans-disciplinarity); how integrated that interdisciplinarity is; the nature and fidelity of the intervention; which outcomes are being measured; how those outcomes are measured and how they are analysed. Unsurprisingly, these studies reach different conclusions. In this situation it is very problematic to integrate these studies and synthesise their findings. In particular, a valid meta-analysis seems particularly unlikely, even of that small percentage of papers which use similar measurement techniques.

However, following the tentative conclusions to the 'review of reviews' and considering these larger studies, we hypothesise that significant effects are less likely (i) for traditional 'standard' outcomes than affective (and perhaps problem solving) outcomes; (ii) for short term than sustained interventions; and (iii) for perceptions of students than teachers'. Even so, this feels like a leap in the dark: we offer these hypotheses as a possibility for informing large scale programme evaluation research in future.

2.3.4 On a Scale of 1 to 5, How Integrated Are You?

There are three relatively equal categories of integration found in our sample of empirical studies. The first is where either (a) mathematics appears in another curriculum subject, or (b) where another subject appears within the mathematics classroom. The second category can loosely be termed thematic integration, where mathematics and other subjects come together around a particular topic or theme, while each retains their disciplinary nature. The third occurs where we have problem- or project-based work where the emphasis is less on bringing subjects together and more on the particular problem or project. These forms of integration also resonate to some extent with notions of mono, multi, inter and trans-disciplinary working. Of course, these categories can overlap to some extent, as we shall see.

Mainly mono-disciplinary. Mathematics in other subjects may be: for the benefit of learning mathematics (e.g. Stone 2007); for the benefit of learning the other subject, using mathematics as a tool or a generalisation (e.g. Andersen 2007); or, truly integrative (e.g. Munier and Merle 2009). It may even potentially transcend the disciplines involved, even if only momentarily. Similarly, other subjects may be brought in to the mathematics classroom, such as history (Bellomo and Wertheimer 2010), literacy (Bintz et al. 2011) or social justice (Bond and Chernoff 2015). Again, although a disciplinary structure remains clear in such cases, the potential exists to vary from being mathematically distinct and mono-disciplinary (e.g. providing motivation, concreteness for the formal mathematics, or wider aspects of mathematical concepts) through to transdisciplinary, when a new 'mathematical' concept might be created to make a problem context malleable.

For example, any particular question within such a lesson may be a genuinely non-disciplinary problem and thus may, at least briefly, break free from typical school mathematics activity. In data handling for instance, we might see a measure of central tendency being created that involves a combination of elimination of incredible outliers with mean average of the credible data points.

Mainly multidisciplinary. Thematic approaches are traditionally seen as multi-disciplinary rather than truly integrative or interdisciplinary (e.g. Carrier et al. 2011, p. 425). A typical example uses a topic such as 'Geography of the Desert' (Patterson and Vetters 1992), where 'the teaching of all subjects—reading, writing, spelling, math, social studies, sciences, and art—centers around the desert'. The connection between subjects can also be a common element between them, such as speed distance and time in mathematics and physics (Ríordáin et al. 2016), or a non-school subject from outside of school (e.g. engineering, Rockland et al. 2010) can provide the focus in which other subjects come together. In such activities, different subjects may begin to interpenetrate and integrate. For example, a thematic unit of 'Mapping: A key to Understanding Our World' was implemented in the third grade of an elementary school (Lonning et al. 1998). This theme-based, interdisciplinary and integrated unit, allowed for connections to be made between topics in mathematics, science and social studies. In social studies, students were introduced to map skills, and city- and state-wide economic issues; in science, rocks and minerals was the focus; and for mathematics, it was patterning, fractions, estimations, classification, area and perimeter. Activities involved students studying the people and places in the city and how the geology and topography of the city impact on settlement and economic development. Mathematics was used as part of the study of rocks and minerals, especially rock patterns, estimation of size and classifications. Science and mathematics becomes a continuum, and integration is argued to help create a balance between both disciplines as activities were shaped by the context while also respecting their individual curricula goals and objectives.

Such perspectives and experiences lead to arguments that interdisciplinary approaches should be centred on *themes* (Ackerman 1989; Jacobs 1989). A theme is a 'topic, concept, problem or issue providing both focus and organising frame-work that guide the development and implementation of a cohesive, interrelated series of lesson or activities' (Lonning et al. 1998, p. 312). Themes have to be well conceived, providing 'a metacognitive bonus—a "powerful idea", a cross-cutting idea, a perspective on perspective taking—that may be of great value' (Ackerman 1989, p. 29). Themes should fulfil three criteria: (a) concepts should be appropriate and important to the individual disciplines, (b) interdisciplinary/integrated instruction should enhance the learning of the concepts and (c) the theme should provide a lens to recognize and understand larger issues and go beyond subject disciplines. Others have argued that using the term 'theme' in this sense just adds to terminological confusion associated with integration and interdisciplinarity (Davison et al. 1995). However, this may necessarily flow from the complexity and variable nature of the activities that come under any particular term.

Mainly interdisciplinary. The key factor in distinguishing between multi-disciplinary and trans-disciplinary perhaps is the balance between (a) the importance of the connecting or unifying issue/subject and (b) maintenance of disciplinary factors such as integrity or rigour. If the theme can rise in importance above issues such as disciplinary assessment or disciplinary timetabling structure, however temporarily, then integration is more likely to occur so that activity is trans-disciplinary. Some empirical examples found in the literature claim to do just that, as the starting point is either a problem or a project, separated from the day-to-day structures of the disciplines. Problem- or project-based work may be easier in primary schools due to organisational structure and rigour of assessment; and there are many examples of this (e.g. Cavin et al. 2014). Within secondary education, external spaces are often helpful (e.g. museums, de Freitas and Bentley 2012), but integration can also be possible within regular school structures (e.g. Ertmer et al. 2014). Some school systems—e.g. in Japan and Canada—have carved out more permanent spaces for interdisciplinary project work in the curriculum at different stages of the school curriculum (cf. Howes et al. 2013). To sustain such attempts however requires moving beyond the obstacles that work against interdisciplinarity, such as the current dominant forms of high-stakes assessment.

2.3.5 Concluding Discussion

This review shows that interdisciplinarity in mathematics education is a relatively under-developed research subfield. The existing literature suggests that there are sometimes beneficial outcomes of interdisciplinary working in integrated curricula, often involving projects. These outcomes emphasise motivational, affective and problem-solving learning outcomes, and perhaps better understandings of what a discipline is, and so how different disciplines can contribute to useful activity. Such suggestions remain contested. Importantly, however, progress in the field is hampered by (a) a lack of clarity and consensus about concepts of disciplinarity, and about how to adequately describe 'interdisciplinary' interventions and programmes; (b) lack of consistency about learning outcomes and how they can be identified and measured; and (c) lack of depth and breadth of research on which future work can build. We hope this gap in the knowledge base of the field will help future researchers place their work constructively.

2.4 Interdisciplinary Teaching and Learning in School—Case Studies

2.4.1 Introduction

Interdisciplinary teaching and learning in school is not a new pedagogical reform, but was promoted a long time ago. The advantages concerning students' understanding of content from different perspectives became evident through many cross discipline teaching experiments (Klafki 1998) and recently, as suggested above, some research studies have begun to produce empirical evidence that promotes these ways of teaching and learning.

In Sect. 2.2 the history and theoretical grounding of interdisciplinarity and interdisciplinary mathematics were described and it became clear how broad this field is from any perspective one takes. Roth's (2014) definition is pragmatic: "Interdisciplinarity denotes the fact, quality, or condition of two or more academic fields or branches of learning. Interdisciplinarity projects tend to cross the traditional boundaries between academic disciplines" (p. 317). In particular, when speaking about interdisciplinary mathematics teaching and learning, Science, Technology, Engineering and Mathematics (STEM) education became a synonym for this "two or more academic fields" not only in the sense that STEM should be fostered more in general, but also in an interdisciplinary manner within one lesson, course or integrated programme.

Promoting STEM education as an integrated curriculum is now a central aspect of educational policy in many countries worldwide, rhetorically in order to prepare students for a more advanced scientific and technological society. The implementation of this demand at the school and university level is still a challenge, and well-grounded practice as well as research is limited in all countries. For example a recently published report of the status quo of STEM-education in Europe (Galev 2015), investigating teachers, students and experts from industry, showed that there has been progress, but STEM-education in school is mainly taught from a more theory-orientated than practice-orientated perspective. Furthermore, teachers and experts of the leading STEM-countries in Europe, like Germany or the United Kingdom, claim that educational policy concerning STEM has not provided expected results, as in many places prospective improvements in STEM take up have not materialised.

However, interdisciplinary mathematics learning and teaching is not, or should not be, limited to the "STE", and might include many other disciplines across the curriculum. It often depends on teacher's knowledge and preferences for a second discipline to cross "traditional boundaries", whether into the Arts, as in STE(A)M, or into humanities or sport. In Germany, for instance, secondary school teachers have to study two school subjects at university to be employed as schoolteachers later on, and every possible combination is allowed! Some beginning mathematics teachers like to choose as a second subject sports or arts, and others not only one of the natural or social sciences. When teaching two subjects to the same class, the idea of, and implementing, transfer across the disciplines is obvious and teachers do this more or less explicitly for their students, even within the traditional structural arrangements of their disciplinary teaching.

Nevertheless interdisciplinary learning and teaching of mathematics requires on the one hand, well-prepared teachers, and on the other hand, adequate teaching materials for every-day lessons in school. Thus, exemplary lesson-units that show how interdisciplinary mathematics can be taught and learnt, build a basis for the teacher's implementation and the development of their own lesson plans. Existing case studies in this field that show, for example, student's views, motivations or performance, while learning in interdisciplinary lessons, are very helpful.

In this chapter the following two case studies give an insight into how disciplines can fit well with mathematics and what might be involved to combine them effectively.

2.4.2 Case Study: Stop Eating to Lose Weight?

Background information and setting. As part of their engagement in two research projects (namely: Mascil, www.mascil-project.eu, and Tomes, www.tomes-project.eu) a group of researchers, teachers, parents, and teacher trainers collabourated together to develop a set of learning activities for the interdisciplinary teaching and learning of elementary statistics. One of the learning activities that was developed, targeted 11–12 year olds, focused on the balance between nutrition and physical activity for a healthy life. Here the other disciplines can be said to be biological science, in the field of health education and nutrition. The activity required students (and parents) to actively participate in the collection, presentation and interpretation of data regarding their nutrition and exercise habits. Students then participated in decision-making processes based on conclusions drawn from the analysis of data (using various representations, like frequency tables, bar charts and pie charts). Through these tasks, students have opportunities to examine, with the aid of spreadsheets and an applet, the variables that may affect the amount of energy intake on a daily basis (e.g. height, mass, age) and suggest specific diet and exercise plans, always taking into consideration the need for balancing the two.

In a period of two months, teachers and parents participated in a number of workshops, delivered by the teacher trainers and researchers, on inquiry based interdisciplinary teaching and learning, and on mathematical modelling and problem solving. Over the course of the next two months, teachers worked collaboratively to develop their own lesson plans and activities, and to implement the activities in their classrooms. Participants met weekly, communicated via email and via a blog designed for the projects, to develop four 80-minute lessons for the *Nutrition and Physical Activity* modelling unit. Further, an interactive applet was also designed to support and facilitate students' work in the activity.

The activity was at first pilot tested in two classes in each one of the three participating schools. Teachers, teacher trainers, parents, and the research team observed each lesson, debriefed and analysed teacher approaches and methods, student work immediately following each observed lesson, and reflected on their understandings throughout the process. Following each lesson implementation individual interviews with the teachers implemented the lesson, and a group interview with all teachers took place. Revised activities were then tried out in other classrooms, followed by interviews, and final modifications in the activities.

The interdisciplinary modelling activity on nutrition and physical activity. The first component of the activity presented the case of Mary, a 14 year old girl who cannot fit into her favourite clothes. The students then considered the general question, "Is stopping eating the right method to lose weight?" Students quickly realized that the question needed to be refined in order to answer it meaningfully,

3. (a) Fill in the following frequency table regarding the habits registered on the questionnaire.

EATING HABITS

FOOD	TALLY MARKS	FREQUENCY
Protein (legumes, fish, chicken, pork, lamb-beef, deli meat products)		
Carbohydrates (bread, pasta, rice)		
Dairies (milk, dairy products)		
Fruit		

Fig. 3 The table students completed after collecting their data

and statistically correctly. On refining the question in their own way, students acknowledged that data on nutrition, and also on physical activity are needed. Following a class discussion on how food consumption is calculated, students agreed to work with their parents at home to anonymously complete a questionnaire regarding their nutrition and physical activity, over a period of one week.

Working with their own data, each group of students summarized their results, by categorizing their data into the different food categories (e.g. protein, carbohydrates, dairy products, fruits, vegetables, sweets, etc.), and by discussing the advantages and disadvantages of each food category (Fig. 3).

The students then used a spreadsheet software to enter their data and they used spreadsheet functions for calculating the sum and average of their entries. In the whole class discussion that followed student groups compared their results in terms of frequency (how often they have food from the different categories), and in terms of quality (which categories are healthier than others). Discussion also focused on first possible suggestions for Mary (e.g. which food might she best 'stop eating', not that frequent, etc.) Students had also opportunities to explore the representation possibilities of the software to generate more detailed representations (see Fig. 4).

On completion of their representations, the students were to respond to the questions, "What does your representation tell you? Are there significant differences between your groups? How does it help to answer the question about Mary's diet? Are more data needed? Why do you think more data are needed?"

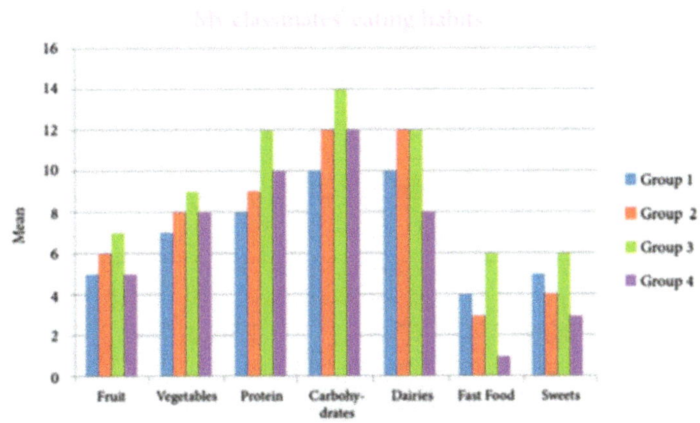

Fig. 4 A bar chart presenting the results for four groups of students

The second component of the activity required students to identify which factors determine a person's 'ideal' daily calorie intake (age, gender, height and body mass). Students worked on analyzing some tables and graphs, via applet software designed for the purposes of the project. Using the applet, students could work on various variables (age, height, mass, level of physical activity, and sports), trying to identify which factors were significant and how they determine a person's ideal daily calorie intake (see Fig. 5).

The third part of the activity commenced by a student statement, indicating that since he likes chocolates, he could just eat nine chocolates to meet his daily calorie needs. Students had to comment on that statement and then worked with a graph presenting the recommended daily intake of servings from each food group for both girls and boys. Students recorded their responses to the following: "From which food category should you consume most in your daily servings? Should boys or girls receive more carbohydrates?" Students then moved to the software, to suggest

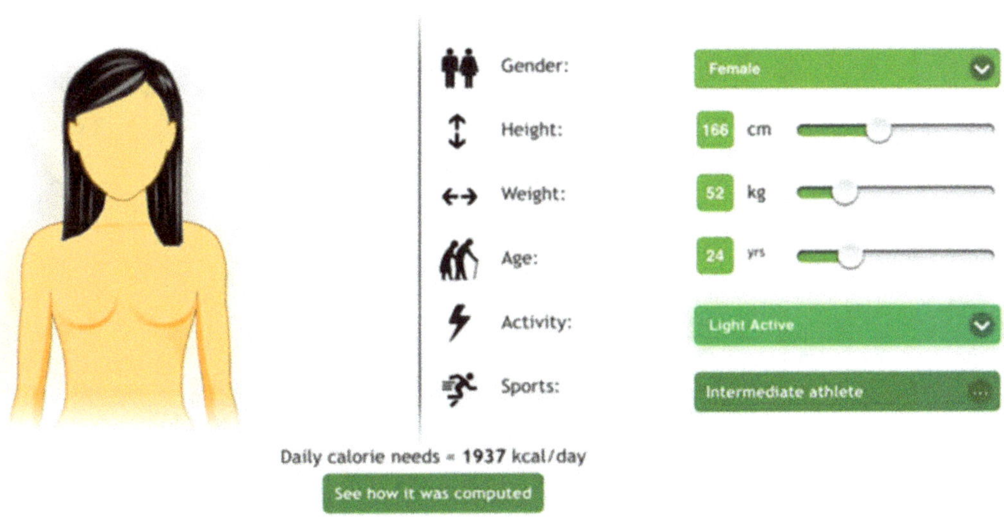

Fig. 5 Applet's screen for selecting the person characteristics

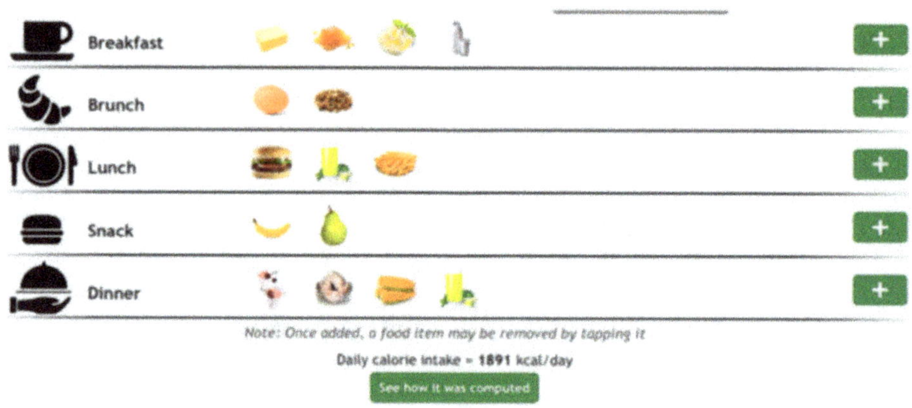

Fig. 6 Creating a person's diet

a balanced diet plan for a day, taking also into consideration the daily amount of energy that a person needs. Students could use the provided 'food database' for creating the person's diet for a day (see Fig. 6) and then explore the appropriateness of the diet with regards to the calories taken and the food categories (see Fig. 7). After completing the tasks and sharing their results in whole class discussion,

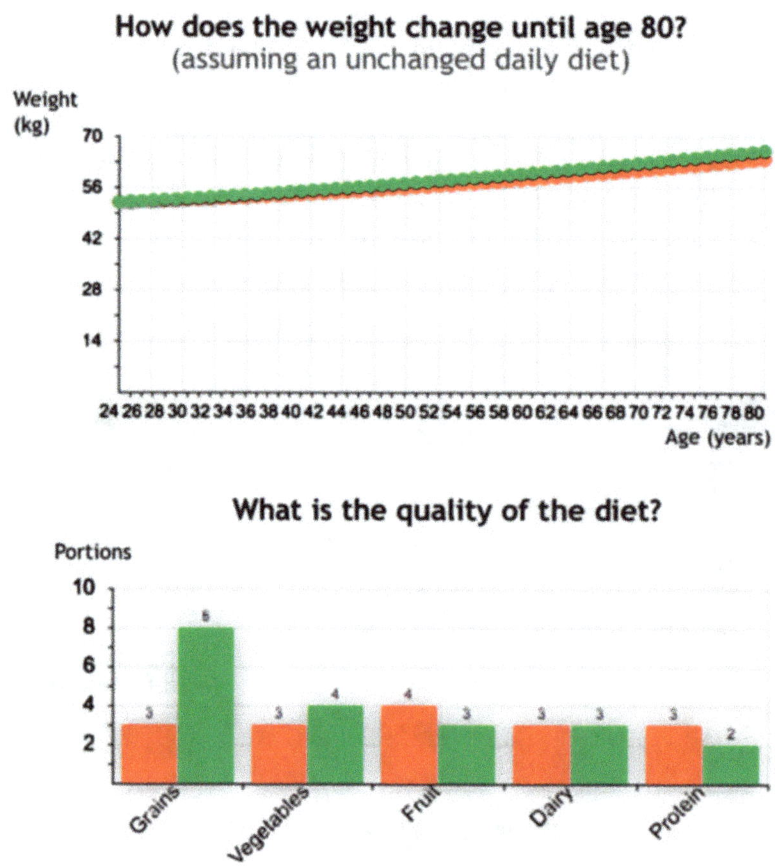

Fig. 7 Examining the appropriateness of a diet in terms of calories and food quality

students then moved to the last part of the task, in which they designed their own balanced nutrition and physical activity case.

Students finally returned to the first question of the activity (Is stopping eating the right method to lose weight?), and documented their answers. Groups of students shared their conclusions with the class, indicating the data they used, their strategies for analysing their data, and how certain they felt about their conclusion.

Reflections on the case study. The interdisciplinary approach reported here focused on the role of the bridging concept, i.e. the balance between calorie intake and exercise. The quite complex real problem setting provided opportunities to explore important concepts from mathematics and science, within a thematic and problem solving context that 'subsumed' the disciplines. Results revealed that teachers found this important for themselves as teachers, since it provided them with a new way of thinking and working. It was also important for students because it provided a new framework in which students focused on solving a real problem, by appropriately studying and using concepts from different school subjects, or disciplines. There was significant potential for transdisciplinarity and meta-disciplinary learning here, as learners could be critical about the lack of relevance of the statistics they were normally expected to calculate in this context of 'real activity' (e.g. 'what is the point of this 'average'?', 'how this is related to the problem and how can it help us answer our question?').

With regards to task design, teachers' and parents' engagement in the design of the activity really improved motivation, and supported the interdisciplinary and real-world nature of the activity. Further, the use of a student-related scenario and the use of students' own real data also contributed to students' motivation, and their willingness to reflect both on the problem setting, but also on their own diet and exercise behaviour.

2.4.3 Case Study: The Successful Student-STEM Programme

Background. This case study is of a well-funded programme across Geelong, a regional city within one hour's drive from Melbourne, the state capital of Victoria, Australia. Situated on Corio Bay, that is part of the much larger Port Phillip Bay, Geelong has a long history as a significant port and industrial city. However, there have been recent changes in the economic climate of Geelong, as the major manufacturing industries (an aluminium smelter, and the Ford Motor car plant) have closed down, and the area has given way to a new, more knowledge-based economy. Funding of $11 AUD million, from the Victorian state government, is key to supporting this transition by focussing on skills development, workforce participation, and education. Specifically, the focus is based on the belief that the future prosperity of the Geelong economy will be underpinned by higher levels of participation and engagement in Science, Technology, Engineering, and Mathematics (STEM) educational pathways. Further, it is believed that developing STEM capability is helping transform it into a technologically advanced and knowledge rich economy.

Skilling the Bay was established in 2011 in response to the Victorian state government's funding initiative, and Deakin University, as a major employer in Geelong, now works in partnership with *Skilling the Bay* to implement a number of programmes. The 'Successful Student—STEM Program' (SS-STEM) programme is one of the initiatives under the *Skilling the Bay* programme, which runs until 2017.

This programme is one of four, aimed at the goal of raising educational attainment and participation levels of students, and addressing the trend of decreasing student participation in STEM at senior levels of secondary schooling. (Data from the Victorian Curriculum and Assessment Authority (VCAA) has shown that Geelong student participation in many senior sciences and mathematics subjects is below metropolitan and state averages.)

The SS-STEM programme. The developers of the SS-STEM programme have a vision of STEM education that goes beyond inter-disciplinary notions. While recognizing the four disciplinary roots of STEM as important, they see the focus of STEM education as being on developing the trunk of the STEM education tree. Thus, the roots support the trunk of the plant, but are no longer visible within it. This metaphor is used to make the point about interdisciplinarity, and transdisciplinarity introduced above more concretely. While this is an ambitious vision, the SS-STEM programme developers believe that this approach will ultimately change teacher practice from the current subject-matter 'silos' to a more coherent and fused perspective.

The programme was developed, and is being implemented, by a team of researchers from the School of Education at Deakin University in partnership with schools and teachers. The initial SS-STEM programme was designed to increase teacher capability and promote student awareness of STEM-related pathways to further study and employment. Although the initial programme funding was to support such a programme, it was later increased to support an additional school-industry partnership programme as well as a STEM Education conference.

The programme involves 10 secondary schools from the Geelong region, focussing explicitly on Year 7 and 8, as these year levels are crucial in laying the foundation for student interests in STEM. Three teachers, from each of the 10 partner schools, are committed to professional development for two-and-a-half years. These teachers may be teachers of mathematics, science, or technology, or teachers in positions of leadership, who can support the change process of the three teachers and within the partner school generally.

Within these two-and-a-half years, teachers undergo three Intensive Professional Development (IPD) sequences focussing on building teachers' knowledge of STEM practices and pedagogies. They then plan and implement a STEM initiative within their school. Such initiatives, thus far, have been units of work, a learning sequence, or a programming structure, that incorporates some interdisciplinary STEM practices and pedagogies promoted through the IPD sequences. In addition, a Deakin Project Officer works with schools to support their developing practice. These IPD sequences, and the continuing support from the university, are critical to supporting each school's approach to STEM innovation.

Schools were required to decide on their own vision for improving STEM. This vision could involve subject-specific innovations (e.g. focussing on mathematics or science only), innovations requiring integration of subjects (e.g. different models of developing activities that involve teaching in science and mathematics), or innovation across a suite of subjects (e.g. promoting particular STEM pedagogies, such as design-based learning, across mathematics and science). However the school vision for STEM was expected to extend beyond the SS-STEM participating teachers. The intention is that, as the programme progresses, the teachers will focus on not only their own development, but also act as change agents in their school to lead sustainable STEM innovation.

One example of a school initiative was the use of a design-based challenge, combining mathematics and science, focussed on simple machines and forces. The STEM practices addressed were creativity, problem-solving, applying tools to a new context, working in teams and communicating findings. Other schools also used design-based challenges as a basis for their STEM initiatives, suggesting that Design Technology may be a vehicle for STEM programmes more generally.

A STEM Teacher network has been established, by the SS-STEM team, to extend support to SS-STEM and non-SS-STEM schools in and around Geelong. Network meetings are held at least once a school term in conjunction with BioLab (a State-funded science centre) in Geelong. BioLab is a cutting edge facility that runs programmes for P-12 students, so is well resourced to illustrate real world problem solving using the latest technologies.

In order to make these meetings accessible and relevant to teachers from the different STEM subjects, the meetings focus on cross-disciplinary themes and complement the professional development offered to SS-STEM teachers. These themes include:

- Students collecting and using real data to inform a problem;
- Challenging students with design-based learning;
- Training students to use appropriate tools, strategically;
- Engaging students in 21st century technological and industrial contexts;
- Using student-generated representations for concept development; and
- Encouraging students to use evidence to justify decisions or construct viable arguments.

In order to support teacher development, the programme uses students, from Deakin's science, engineering and IT faculties, to act as ambassadors for STEM in their chosen career, and to assist teachers with implementing STEM initiatives and projects. In 2015 four student ambassadors went to three of the schools and were well received by the teachers and the school students.

A STEM into Industry programme is currently underway (2016) to assist schools to use industry links in their STEM units. The Australian Industry Group (AIG) is assisting with establishing relationships between schools and Geelong-based industries. The intention is that these relationships will enable development of

STEM curricula that are tailored to industry and school needs, and that illustrate the direct relationships between education, industry, and future employment.

The STEM Education Conference, October 2016, has an open invitation to teachers, academics, researchers and curriculum developers with an interest in STEM education in schools. There will also be visits to the SS-STEM partner schools on offer as part of the conference. This conference will provide an important opportunity for teachers from SS-STEM schools to showcase their innovations, and for all teachers, especially in the Geelong region, to see and hear about possibilities for STEM education.

Changes in teachers' practice. An important aspect of the SS-STEM programme is the tracking of teacher perspectives of their practice. This aspect of the programme uses Component Mapping (Groves et al. 2007) in which teachers are asked to rate themselves, on a five-point Likert scale, on a series of statements about their current pedagogy. Early results from the end of the first year of the programme show a small, but positive change. The Component Mapping for 2015 average score was 39.37 (SD = 13.28) and in 2016 was 39.36 (SD = 11.54), showing that the distribution of teachers' former practice has changed after exposure to inter-disciplinary pedagogy, but, significant gains have not yet been made. The reduction in the Standard Deviation might suggest that the teachers, as a group, have moved towards a more common perception of their pedagogy, possibly due to the professional development offered to SS-STEM teachers.

In conclusion, this case study (in progress) shows some signs of what can be achieved in terms of pedagogic transformation given conditions: (i) the political will and funding; (ii) Intensive Professional Development of multi-disciplinary groups of teachers; and (iii) with IDP sequences based on interdisciplinary and transdisciplinary principles (e.g. through thematic, and industrial problem solving).

2.4.4 Discussion

These two case studies illustrate what inter-disciplinary mathematics can be, and how it is being realized in schools and by teachers in diverse ways. The first case study showed a link between disciplines being made, but which teachers cannot be expected to know or understand if it is not made explicit, and supported by curriculum development projects. Simultaneously, it showed how students were engaged in inter-disciplinary practice based in their 'real' world of experience, i.e. outside academe/schooling, involving the support of the community, in this case the parents and families of the learners. In this case it seemed clear that Mathematics was the 'lead' discipline, but not necessarily the 'leading activity'. That is, the knowledge that counted in solving the problem was not only mathematics, but at least as important was the information about food nutrition.

In the second case study, through a large-scale multi-site Professional Development programme, teachers are being provided with long term support and professional development around IdME, and were beginning to see that their current pedagogy was not good inter-disciplinary practice. This disturbance in the field is an important factor in preparing for change in teaching (and teachers). Additionally, the case stresses the role of political and economic factors in setting a STEM agenda; being in an economically depressed region, political factors allow the traditional norms to be challenged, not least in providing essential resources for such a major initiative.

We are conscious in reporting only two cases (one still under way, the other only light-touch research) that we are scratching the surface of the activities that are being studied. One imagines that a synthesis of many such case studies would be needed to identify common success and risk factors; and then even qualitative evaluation case studies that develop theory are hugely expensive and time consuming (compared to resources available). Clearly conditions are important: the challenges for 'teachers working together' across disciplinary boundaries are very different in Primary and secondary for instance. Additionally, long term change involves consistency, freedom to develop and assess the curriculum in new ways etc. (Howes et al. 2013).

The two cases do also raise some important questions about how research should contribute to and benefit from all this work. How is success to be defined, evaluated, assessed? Can there be consistency for the purposes of evaluation? Who has agency in the development process and what are the necessary conditions for resources?

Undoubtedly there is further work to be done, but such cases provide ideas for re-thinking curriculum and pedagogy, and set possible yard-sticks for evaluation of developments of this kind.

3 Summary and Looking Ahead

In conclusion, we have begun the work of this study group in three ways. First we have posed the need for conceptual clarity, and situated the different conceptions of interdisciplinarity in a social, cultural, and historical account of how disciplines in general have arisen, and how they have functioned to nurture the growing division of labour in society. It also showed how disciplinary discourses and practices invoke professional identity and power relations that need to be understood. Perhaps also we may understand the dysfunctionality of our multidisciplinary Towers of Babel, especially in the academic, scientific world. The widely accepted terminology of mono-, multi-, inter-, trans-, and meta-disciplinarity was introduced (but not uncritically so!).

Second, we have begun a State of the Art literature survey of the empirical research, drawing some hesitant inferences for future research. This hesitation is partly due to weaknesses in the quality and quantity of the research base, and partly because we have yet to complete this review work. Our survey was able to illustrate the type of work and outcomes the extant literature has produced, and point to some substantive features that reviews of research and the research generally agrees on.

To counter this we have shown some key limitations in the research, substantively and methodologically.

Thirdly, we have offered some Research and Development case studies that were well founded and that illustrate the kind of practical development work being undertaken these days.

There are so many limitations to this work for researchers to fill we confidently expect this topic to provide work for some decades yet: we have no more than touched on whole research fields such as the History of Mathematics, Statistics, Quantitative methods in undergraduate social sciences, interdisciplinarity outside academe, and so on. We have tried to make sure the reader was made aware of many of these limitations as we progressed in the text, but there are too many to do so comprehensively without being boring.

Outcomes from this study include, in summary:

- Theoretical roots of 'disciplines' in a social, cultural and historical analysis of their practices and discourses.
- A continuum of disciplinarity and its concepts, including mono-disciplinarity, multi-disciplinarity, interdisciplinarity, transdisciplinarity, and meta-disciplinarity; and a critique of this continuum.
- A literature survey and preliminary review of empirical research literature; with conclusions for future research.
- Case studies of Research and Development of interdisciplinarity in schools in the field.

References

Ackerman, D. B. (1989). *Intellectual and practical criteria for successful curriculum integration* (pp. 25–37). Interdisciplinary curriculum: Design and Implementation.

Alspaugh, J. W., & Harting, R. D. (1998). Interdisciplinary team teaching versus departmentalization in middle schools. *Research in Middle Level Education Quarterly, 21*(4), 31–42.

Andersen, J. (2007). Enriching the teaching of biology with mathematical concepts. *American Biology Teacher, 69*(4), 205–209.

Becker, K., & Park, K. (2011). Effects of integrative approaches among science, technology, engineering, and mathematics (STEM) subjects on students' learning: A preliminary meta-analysis. *Journal of STEM Education: Innovations & Research, 12*(5), 23–37.

Bellomo, C., & Wertheimer, C. (2010). A discussion and experiment on incorporating history into the mathematics classroom. *Journal of College Teaching & Learning, 7*(4), 19–24.

Berlin, D. F., & Lee, H. (2005). Integrating science and mathematics education: Historical analysis. *School Science and Mathematics, 105*(1), 15–24.

Bintz, W. P., Moore, S. D., Wright, P., & Dempsey, L. (2011). Using literature to teach measurement. *Reading Teacher, 65*(1), 58–70.

Bond, G., & Chernoff, E. J. (2015). Mathematics and social justice: A symbiotic pedagogy. *Journal of Urban Mathematics Education, 8*(1), 24–30.

Bourdieu, P. (1984). *Homo academicus [Homo academicus]*. Paris: Les Éditions de Minuit.

Bourdieu, P. (1992). The practice of reflexive sociology (the Paris workshop). In P. Bourdieu & L. J. D. Wacquant (Eds.), *An invitation to reflexive sociology* (pp. 217–260). Chicago, IL: The University of Chicago Press.

Bourdieu, P. (2000). *Pascalian meditations*. Stanford, CA: Stanford University Press.

Boyd, B., Dunlop, A.W., Mitchell, J., Logue, J., Gavienas, E., Seagraves, L., Clinton, C., & Deuchar, R. (2007). *Curriculum architecture-a literature review*. Report for The Scottish Executive Education Department to inform the implementation of A Curriculum for Excellence, 2–120.

Carrier, S., Wiebe, E. N., Gray, P., & Teachout, D. (2011). BioMusic in the Classroom: Interdisciplinary elementary science and music curriculum development. *School Science and Mathematics, 111*(8), 425–434.

Cavin, A., Elfer, C. J., & Roberts, S. L. (2014). iGardening: Integrated activities for teaching in the common core era. *Social Studies and the Young Learner, 26*(4), 5–9.

Collen, A. (2002). Disciplinarity in the pursuit of knowledge. In G. Minati & E. Pessa (Eds.), *Emergence in complex, cognitive, social, and biological systems* (pp. 285–296). New York, NY: Springer.

Czerniak, C. M., Weber, W. B., Sandmann, A., & Ahern, J. (1999). A literature review of science and mathematics integration. *School Science and Mathematics, 99*(8), 421–430.

d'Ambrosio, U. (1990). Literacy, matheracy, and technocracy: A trivium for today. *Mathematical Thinking and Learning, 1*, 131–153.

Davison, D. M., Miller, K. W., & Methany, D. L. (1995). What does integration of science and mathematics really mean? *School Science and Mathematics, 95*(5), 226–230.

de Freitas, E., & Bentley, S. J. (2012). Material encounters with mathematics: The case for museum based cross-curricular integration. *International Journal of Educational Research, 55*, 36–47.

Delamont, S., Atkinson, P. A., & Odette, P. (2000). *The doctoral experience: success and failure in graduate school*. New York: Falmer.

Dorn, R. I., Douglass, J., Ekiss, G. O., Trapido-Lurie, B., Comeauz, M., Mings, R., et al. (2005). Learning geography promotes learning math: Results and implications of Arizona's GeoMath grade K-8 program. *Journal of Geography, 104*(4), 151–160.

Durkheim, É. (1893). *De la division du travail social [division of social labour]*. Paris: Felix Alcan.

Durkheim, É. (1915). *Elementary forms of religious life*. London: George Allen & Unwin.

Ehn, P., & Kyng, M. (1991). Cardboard computers: Mocking-it-up or hands-on the future. In J. Greenbaum & M. Kyng (Eds.), *Design at work: Cooperative design of computer systems* (pp. 169–195). Hillsdale, NJ: Lawrence Erlbaum Associates.

Ertmer, P. A., Schlosser, S., Clase, K., & Adedokun, O. (2014). The grand challenge: Helping teachers learn/teach cutting-edge science via a PBL approach. *Interdisciplinary Journal of Problem-based Learning, 8*(1), 4–20.

Foucault, M. (1970). *The order of things*. New York, NY: Random House.

Foucault, M. (1972). *Archeology of knowledge and the discourse on language*. New York, NY: Pantheon Books.

Foucault, M. (1978). *Discipline and punish: The birth of the prison*. New York, NY: Vintage Books.

Foucault, M. (1988). *Madness and civilization: A history of insanity in the age of reason*. New York, NY: Vintage Books.

Frodeman, R. (2010). Introduction. In R. Frodeman, J. T. Klein, & C. Mitcham (Eds.), *The Oxford handbook of interdisciplinarity* (pp. xxix–xxxix). Oxford: Oxford University Press.

Fuller, S. (1991). Disciplinary boundaries and the rhetoric of the social sciences. *Poetics Today, 12*, 301–325.

Galev, T. (2015). *Status Quo der MINT-Bildung in Europa: Ergebnisse der MARCH Studie*. European Union: Projekt Lebenslanges Lernen.

Groves, S., Doig, B., & Tytler, R. (2007). Fun or profit: Primary students' perceptions of mathematics and science. In J. Novotna & H. Moraova (Eds.), *Proceedings of the international symposium elementary mathematics teaching: Approaches to teaching mathematics at the elementary level* (pp. 126–133). Prague: Charles University.

Henderson, K. (1991). Flexible sketches and inflexible data bases: Visual communication, conscription devices, and boundary objects in design engineering. *Science, Technology and Human Values, 16*, 448–473.

Hicks, D. (1992). Instrumentation, interdisciplinary knowledge, and research performance in spin glass and superfluid helium three. *Science, Technology and Human Values, 17*, 180–204.

Hine, C. (1995). Representations of information technology in disciplinary development: Disappearing plants and invisible networks. *Science, Technology and Human Values, 20*, 65–85.

Honey, M., Pearson, G., & Schweingruber, H. (Eds.) (2014). *STEM integration in K-12 education: Status, prospects, and an agenda for research*. National Academies Press.

Howes, A., Kaneva, D., Swanson, D., & Williams, J. (2013). Re-envisioning STEM education: Curriculum, assessment and integrated, interdisciplinary studies. A report for the Royal Society. https://royalsociety.org/~/media/education/policy/vision/reports/ev-2-vision-research-report-20140624.pdf. Accessed 16 May 2016

Hurley, M. M. (2001). Reviewing integrated science and mathematics: The search for evidence and definitions from new perspectives. *School Science and Mathematics, 101*(5), 259–268.

Jacobs, H. H. (1989). *Interdisciplinary curriculum: Design and implementation*. Alexandria, VA: Association for Supervision and Curriculum Development.

Klafki, W. (1998). *Kriterien einer guten Schule*. Marburg.

Klein, J. T. (2010). A taxonomy of interdisciplinarity. In R. Frodeman, J. T. Klein, & M. Mitcham (Eds.), *The Oxford handbook of interdisciplinarity* (pp. 15–30). Oxford: Oxford University Press.

Leont'ev, A. N. (1978). *Activity, consciousness and personality*. Englewood Cliffs, NJ: Prentice Hall.

Lonning, R. A., DeFranco, T. C., & Weinland, T. P. (1998). Development of theme-based, interdisciplinary, integrated curriculum: A theoretical model. *School Science and Mathematics, 98*(6), 312–319.

Lynch, M. (1985). Discipline and the material form of images. *Social Studies of Science, 15*, 37–66.

Mannheim, K. (1956). *Essays on the sociology of culture. Collected works* (Vol. 7). London: Routledge.

Marcovich, A., & Shinn, T. (2011). Where is disciplinarity going? Meeting on the borderland. *Social Science Information, 50*, 582–606.

Marx, K., & Engels, F. (1962). *Werke Band 23* [Works vol. 23]. Berlin: Dietz.

Marx, K., & Engels, F. (1978). *Werke Bd. 3* [Works vol. 3]. Berlin: Dietz.

Miettinen, R. (1998). Object construction and networks in research work: The case of research on cellulose-degrading enzymes. *Social Studies of Science, 28*, 423–463.

Munier, V., & Merle, H. (2009). Interdisciplinary Mathematics-Physics Approaches to Teaching the Concept of Angle in Elementary School. *International Journal of Science Education, 31* (14), 1857–1895.

Onwuegbuzie, A. J., & Wilson, V. A. (2003). Statistics anxiety: Nature, etiology, antecedents, effects, and treatments–a comprehensive review of the literature. *Teaching Higher Education, 8*, 195–209. doi:10.1080/1356251032000052447

Parr, B., Edwards, C. M., & Leising, J. G. (2009). Selected effects of a curriculum integration intervention on the mathematics performance of secondary students enrolled in an agricultural power and technology course: An experimental study. *Journal of Agricultural Education, 50* (1), 57–69.

Patterson, K., & Vetters, L. (1992). Geography of the desert: An interdisciplinary approach. *Journal of Geography, 91*(4), 143.

Ríordáin, M. N., Johnston, J., & Walshe, G. (2016). Making mathematics and science integration happen: key aspects of practice. *International Journal of Mathematical Education in Science and Technology*, *47*(2), 233–255.

Rockland, R., Bloom, D. S., Carpinelli, J., Burr-Alexander, L., Hirsch, L. S., & Kimmel, H. (2010). Advancing the "E" in K-12 STEM education. *Journal of Technology Studies, 36*(1), 53–64.

Roth, W.-M. (2014). Interdisciplinary approaches in mathematics education. In S. Lerman (Ed.), *Encyclopedia of mathematics education* (pp. 647–650). Berlin, Heidelberg: Springer.

Roth, W. M., & Bowen, G. M. (2001). Of disciplined minds and disciplined bodies. *Qualitative Sociology, 24*, 459–481.

Shulman, V., & Armitage, D. (2005). Project discovery: An urban middle school reform effort. *Education and Urban Society, 37*(4), 371–397.

Shumway, D. R., & Messer-Davidow, E. (1991). Disciplinarity: An introduction. *Poetics Today, 12*, 201–225.

Silk, E. M., & Schunn, C. D. (2011). A cognitive perspective on integrated STEM learning. http://www.nae.edu/Projects/iSTEM.aspx. Accessed 16 May 2016

Simmel, G. (1890). *Über soziale Differenzierung: Soziologische und psychologische Untersuchungen [On social differentiation: Sociological and psychological investigations]*. Leipzig: Duncker & Humblot.

St. Clair, B. & Hough, D. L. (1992). Interdisciplinary teaching: A review of the literature. http://files.eric.ed.gov/fulltext/ED373056.pdf. Accessed 14 May 2016

Star, S. L. (1995). The politics of formal representations: Wizards, gurus, and organizational complexity. In S. L. Star (Ed.), *Ecologies of knowledge: Work and politics in science and technology* (pp. 88–118). Albany: State University of New York Press.

Stevens, R., Wineburg, S., Herrenkohl, L. R., & Bell, P. (2005). Comparative understanding of school subjects: Past, present, and future. *Review of Educational Research, 75*(2), 125–157.

Stinson, K., Harkness, S. S., Meyer, H., & Stallworth, J. (2009). Mathematics and science integration: Models and characterizations. *School Science and Mathematics, 109*(3), 153–161.

Stone, J. R. (2007). Making math work. *Principal Leadership, 7*(5), 43–45.

Traweek, S. (1988). *Beamtimes and lifetimes: The world of high energy physicists*. Cambridge, MA: MIT Press.

Turnbull, D. (1993). The ad hoc collective work of building gothic cathedrals with templates, string, and geometry. *Science, Technology and Human Values, 18*, 315–340.

Venville, G., Rennie, L. J., & Wallace, J. (2012). Curriculum integration: Challenging the assumption of school science as powerful knowledge. *Second international handbook of science education* (pp. 737–749). Netherlands: Springer.

Further Reading

This Royal Society report covers much ground in relation to curriculum, assessment and integrated studies:

Howes, A., Kaneva, D., Swanson, D., & Williams, J. (2013). Re-envisioning STEM education: Curriculum, assessment and integrated, interdisciplinary studies. A report for the Royal Society. https://royalsociety.org/~/media/education/policy/vision/reports/ev-2-vision-research-report-20140624.pdf. Accessed 16 May 2016

Terence Tau reveals how *intra*-mathematical problem solving by large teams of mathematicians required certain constraints we associate with inter-disciplinarity (and more!) here: https://www.youtube.com/watch?v=elWIDVI6b18. Accessed 30 May 2016

www.ingramcontent.com/pod-product-compliance
Lightning Source LLC
LaVergne TN
LVHW082240200926
848484LV00008B/2384